A BEACON BIOGRAPHY

Bill Nye

Tamra B. Orr

PURPLE TOAD
PUBLISHING

PURPLE TOAD
PUBLISHING

Printing   1   2   3   4   5   6   7   8   9

A Beacon Biography

Angelina Jolie
Anthony Davis
Big Time Rush
Bill Nye
Cam Newton
Carly Rae Jepsen
Carson Wentz
Chadwick Boseman
Daisy Ridley
Drake
Ed Sheeran
Ellen DeGeneres
Elon Musk
Ezekiel Elliott
Gal Gadot
Harry Styles of One Direction
Jennifer Lawrence

John Boyega
Kevin Durant
Lorde
Malala
Maria von Trapp
Markus "Notch" Persson, Creator of Minecraft
Millie Bobby Brown
Misty Copeland
Mo'ne Davis
Muhammad Ali
Neil deGrasse Tyson
Peyton Manning
Robert Griffin III (RG3)
Stephen Colbert
Stephen Curry
Tom Holland
Zendaya

Publisher's Cataloging-in-Publication Data
Orr, Tamra B.
  Bill Nye, the Science Guy / written by Tamra B. Orr.
    p. cm.
Includes bibliographic references, glossary, and index.
ISBN 9781624693571
1. Nye, Bill. 2. Science television programs--Juvenile literature. 3. Science in mass media--Juvenile literature. 4. Mechanical engineers--United States--Biography--Juvenile literature. I. Series: Beacon biography.
 Q141 2017
 509.2

Library of Congress Control Number: 2017957793

eBook ISBN: 9781624693939

ABOUT THE AUTHOR: Tamra B. Orr is a full-time author living in the Pacific Northwest with her family. She graduated from Ball State University in Muncie, Indiana. She has written more than 500 books about everything from historical events and career choices to controversial issues and celebrity biographies. On those rare occasions that she is not writing a book, she is reading one. She watched Bill Nye on television with her kids and still loves catching him on some of her favorite TV shows.

PUBLISHER'S NOTE: This story has not been authorized or endorsed by Bill Nye.

CONTENTS

Bill Nye has been a favorite guest scientist at places like the Goddard Space Flight Center in Greenbelt, Maryland.

Say the name "Bill Nye" to most people, and they are likely to respond with "The Science Guy!" That is not a surprise. Nye is one of the best-known scientists in the world. He's an inventor, an educator, a science advocate—and a comedian. While there are many reasons he loves science, one of the biggest actually happened years before he was born.

In the 1940s, during World War II, many American soldiers were captured by the Japanese. Some of them were held for years in prisoner of war (POW) camps. One of these POWs was a quartermaster named Edwin Darby Nye, known to his buddies as Ned. Ned was held in a camp for more than four years. Conditions were terrible, and his captors cruel. He needed to do something to keep himself busy—and to stay sane. He began building sundials out of fence posts and pebbles. Sundials are ancient instruments used for telling time. As the sun moves across the sky, the shadows change, making it possible to know the hour of the day by the shadows cast.

Want to make your own sundial? First, find a sunny spot in the lawn or on a sidewalk. Stick a pencil in soft clay, and place it on the ground. As the sun moves across the sky, place a rock or chalk mark for each hour to show where the shadow falls at that time. It may take a few days for you to fill in all hours of the day.

Each day in the camp, Ned would check his sundial to see what time it was. He used the sundial to keep track of the passing of the days, weeks, months—and years.

Finally Ned was released. When he returned home, he worked as a traveling salesman around Washington, D.C., but his interest in sundials kept growing. According to an interview in *The New Yorker*, he would stop by restaurants for his usual sandwich and iced coffee and ask his server, "Know of any sundials around here?" In the end, Ned's passion for sundials resulted in two things. He wrote a book

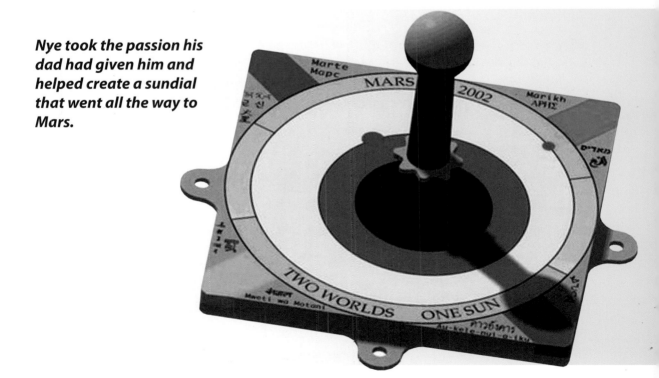

about them, and he passed his excitement for science on to his son, Bill.

"My dad was fascinated with sundials," Bill Nye told *AstroBiology Magazine*. "He photographed hundreds of them and wrote a book about them. He designed a Nye family dial that kept time at our house in Washington, D.C." In the *New Yorker* interview, Nye added, "As a child, I would stand with my father on the beach in Delaware and stare at my shadow as the sun went down, watching it get longer and longer, infinitely long."

Years later, Bill would get the chance to build a sundial that would be flown to Mars. It was a moment Ned would have loved, had he still been alive to see it. He also would have been amazed and proud of the career in science that his son would come to embrace.

Nye took the passion his dad had given him and helped create a sundial that went all the way to Mars.

Nye's mother, Jacqueline (left), was a math whiz. She helped the United States during World War II.

Becoming a Scientist

From the time he was a young boy, William Sanford Nye, or Bill, loved science. Born to Jacqueline and Ned Nye on November 27, 1955, his family would include sister Susan and brother Edward. While Ned was fascinated by sundials, Jacqueline was brilliant at math. During the same war when her husband was taken prisoner, Jacqueline worked for the military as a codebreaker. It was her job to crack the codes used by the Japanese and the Germans. When a reporter at the *Baltimore Sun* asked about his mother, Bill said, "Her influence on me was infinite, immeasurable. She taught me how to cook—and how to make the famous family salad dressing that her grandmother taught her to make—which was pure chemistry. She taught me how to sew, even," he added. "To this day, I still own a sewing machine. And to this day, I can still hear her chanting in my ear: 'Sit up straight! Shoulders back! Now train, train, train! Do it till you get it right!'"

Nye's mind was always working. He wanted to know how his bike wheels worked. How could those flimsy spokes keep the rims from collapsing? "When I was a kid, I thought bicycles were the coolest thing," Nye stated in a video interview with PBS. He wanted to figure out how to make his rubber band–powered airplane turn. Everything was interesting to him. "I watched bumblebees for hours," he told *Popular Science*. "How could such a relatively big animal fly with such relatively small wings?"

The world was full of scientific mysteries like these. Even today, Nye encourages young people to never use the word *bored* or *boring*, especially about science. It says more about the person than about the subject. He told *Popular Science*, "It's hard for me to imagine being 'bored' ever. The world is so exciting and fascinating, yes?"

Nye also loved playing games with flying discs. When he was young, he threw a Pluto Platter, a disc that would come to be known as a Frisbee. "I was comfortable throwing a disc around," he told a reporter at *Ultiworld*. Nye has continued to play some form of Frisbee ever since. "I just wanted to be part of it," he explained. "I understood it instantly."

Even in high school, Nye was crazy about science. He spent hours trying to understand the world around him.

After going to Lafayette Elementary and Alice Deal Junior High, Nye was given a partial scholarship to a private school called the Sidwell Friends School in D.C. He graduated in 1973, and then it was time for college. Nye chose Cornell University in Ithaca, New York. "My parents both worked so hard to get me into Cornell and pay for it, it's amazing," Nye told *Ultiworld*. "It changed my life. Cornell made me who I am."

While Nye was at Cornell, he took an astronomy class with the world-famous astrophysicist Carl Sagan. In a YouTube video recorded by Nye, he said, "His classes were just fantastic." Nye graduated with a Bachelor of Science in Mechanical Engineering.

One of the world's best-known scientists, Carl Sagan, was a huge influence on Nye. Sagan was an astronomer, cosmologist, astrophysicist, author, and teacher.

During his senior year in college, a friend came running to his house to get him to watch the comedian Steve Martin do his routine on television. "He said, 'Look, this guy is just like you,'" Nye told PBS. " 'You're just like this guy. This is what you should be doing.' " Nye just laughed. Once he graduated, he got a job as a mechanical engineer.

However, about a year later, things changed. He was working for Boeing Air in Seattle and decided to enter a Steve Martin look-alike contest. He won! "People wanted me to be Steve Martin at parties," he explained. He was asked to perform comedy routines—and decided to try it. He loved it. "After you get laughs on stage, it's addicting," he admitted to PBS. He starting appearing on the Seattle-based comedy show called *Almost Live!* While on that show, the host, Ross Schafer, asked Nye to do a comedy skit using science. He called him "Bill Nye, the Science Guy" for the first time. And a new career began!

Nye's job as a comedian
was very different from his
job at Boeing.

During the day, Nye enjoyed working at the Boeing Company as an
engineer. While he was there, he invented a complex, high-tech piece
of equipment for the Boeing 747 airplanes. It was one of several
patents that bear his name. The piece is still being used in planes.

In his free time, Nye also enjoyed being a comedian on Seattle's
Almost Live! He also wrote and performed jokes on Seattle's KJR Radio
station. He told *Fast Company*, "I was working on jet navigation
systems . . . during the day, and I'd take a nap and go do stand-up
comedy by night."

By 1991, he had recurring skits on *Almost Live!*, including one as the
Science Guy. He was also a technical adviser on the *Back to the Future*
animated series, appearing in live-action clips at the end of each show.
He did not speak but showed viewers how to do the science that
Christopher Lloyd, or Dr. Emmett Brown, was discussing. His
character wore a bow tie and a lab coat.

Nye liked wearing bow ties and light blue lab coats when he was in
character. The coats made him look more like a scientist. As for the

bow ties? He just likes wearing them. He says he currently has a few hundred of them—and none of them are clip-ons. "Yes, I tie my own ties, my goodness," he told PBS. He also suggests everyone try wearing a bow tie. "Don't take my word for it," he said. "Wear one and see what happens!"

Nye was torn between science and comedy. Was there a way for him to combine them? He gave it a great deal of thought. Finally, he came up with a way to have the best of both worlds. He wanted to teach science to children while also having fun. He decided to start an educational television show about science. He would base it on his Science Guy character.

People grew to recognize Nye from the bow ties he wore. Together with the lab coat, they conveyed the image of a zany scientist.

"I worked at an engineering firm . . . for people obsessed with making a profit every quarter. . . . You cannot advance much with that outlook," he said in an interview with *Popular Science*. "So, I decided to affect the future as much as I could; I shifted my focus to elementary science education." Nye knew just what he wanted to do with his show: he wanted to change the world.

In 1993, he got that chance. The first episode of *Bill Nye, the Science Guy* aired on PBS on September 10. The theme song was definitely different from those of other children's shows. It was loud, with many voices. In the background, singers chanted, "Bill! Bill! Bill!" Over the next five years, it ran for 100 episodes. It won 19 Emmys. The show

Like his mentor Carl Sagan, Nye found a way to bring science home to the family and make it interesting and easy to understand.

was used in science classrooms across the country to help explain topics like digestion, gravity, and space exploration. In lab coat and bow tie, Bill Nye taught children to love science every single week.

Bill Nye, the Science Guy ended on June 20, 1998, but Nye was not finished sharing science with young people. Videos of the entire series were made available for schools to show in classrooms. Meanwhile, in 1998, he played a science teacher in the Disney movie *The Principal Takes a Holiday*. He also appeared in various television shows such as *Numb3rs*, *How to Be a Millionaire*, and *Stargate Atlantis*.

In addition to being on TV, Nye was also inventing things. He created a magnifying glass made with a plastic bag filled with water. He made a device that guides a baseball throw. He even invented a better type of ballet shoe.

While making an episode about bones and muscles for *Bill Nye, the Science Guy*, Nye watched a group of dancers from the Pacific Northwest Ballet Company. He saw that their shoes had blood on them. When dancers would go en pointe, or up on their toes, it would damage their feet.

"These women, they're 22 years old, and they have three or four surgeries already," Nye told *Fast Company*. "They're covering up their scars with makeup. I just got to thinking about it. The toe shoe has not changed in centuries. So I just got to thinking about it."

Nye applied his knowledge of the human foot to his knowledge of physics and created a new type of ballet shoe. It gives more support to the foot, and includes an area that protects the dancer's toes.

From 2000 to 2002, Nye served as the technical expert for the robot competition show called *BattleBots*. The show featured high-tech robots that battle each other to see which is the toughest.

In 2005, Nye was back on television, this time hosting the show *Eyes of Nye*. This show was on deeper science topics for a slightly older audience. For example, one episode explored our car culture, including traffic engineering. Another explored global climate change. The show lasted for one season of thirteen episodes.

Nye also hosted thirteen episodes for Planet Green Network's *Stuff Happens*, the Science Channel's *100 Greatest Discoveries*, and an eight-part miniseries for the Discovery Channel's *Greatest Inventions*. All of these jobs kept Nye busy, but he was about to work on a project that would take him back to his days of being with his dad.

Nye's shows have helped families learn a great deal about science, but it was time for his knowledge to go past the planet Earth.

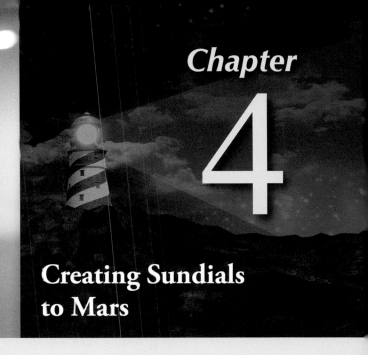

Chapter 4

Creating Sundials to Mars

At the end of the twentieth century, a group of scientists from Cornell University sat around a table, discussing how they were going to design the probes that would be sent to Mars. Some problems were fairly easy to solve. Some were not. One question they were struggling with involved how to measure time on the probe. One Cornell graduate—Bill Nye—had the perfect idea. "C'mon guys, it's got to be a sundial!" he said, according to an interview in *The New Yorker*. The other scientists were not so sure. "Bill, we've got a lot of clocks already, man—it's a space program. *Space program*," Nye recalled them saying.

Nye knew a great deal about sundials, thanks to his dad. He built two MarsDials for the Mars Exploration Rovers. They are small, about the size of a person's hand. Made out of aluminum, they are engraved with the words *Two Worlds; One Sun*, plus the word *Mars* in seventeen languages. According to *Astrobiology Magazine*, alongside some stick figures (referred to as "sticksters") on the sundials, Nye wrote:

This is an artist's illustration of the rover that was sent to Mars in 2003.

People launched this spacecraft from Earth in our year 2003. It arrived on Mars in 2004. We built its instruments to study the Martian environment and to look for signs of water and life. We used this post and these patterns to adjust our cameras and as a sundial to reckon the passage of time. The drawings and words represent the people of Earth. We sent this craft in peace to learn about Mars' past and about our future. To those who visit here, we wish a safe journey and the joy of discovery.

Nye said that he was inspired to write these words because of his former teacher, Carl Sagan. "Professor Sagan asked the class which songs we thought were worthy of being sent out of our Solar System, messages in bottles to be cast into the Cosmic Ocean," Nye told *Astrobiology Magazine*. "The chance that they will ever be found is

Years ago, Sagan posed with a model of the Viking moon lander.

almost unimaginably remote. But, it is also a notion that fills us with wonder and hope. The idea is irresistible—a message from humankind for all time." Sagan went on to tell students about the plaques that were on the side of Voyager spacecraft. "That made quite an impression on me as a student," explained Nye. The chance to make a sundial and send it to Mars "is a thrill I may never match," he said. "I am honored to have been included and given a chance to contribute to space exploration."

Over the years, Nye also kept busy with a number of scientific organizations. For five years, he was part of the Planetary Society, a large space interest group. He is also a member of the Committee for Skeptical Inquiry, a group that promotes science and investigation. In all the roles that Nye has played, his mission has remained the same. As he said in an interview with *Live Science*, it is "to help foster a scientifically literate society, to help people everywhere understand and appreciate the science that makes our world work." That is a life motto that Ned would certainly have been proud to hear.

Saving
the World

Bill Nye is not above stirring up trouble—or reaching out to try
something new.

Science has many topics that are controversial. Nye is not afraid to
state his thoughts and opinions about each one of them. He often
speaks out against people who are anti-vaccination, or those who deny
climate change. In 2012, in response to an interview question, Nye
offended a number of religious people with his statements about
evolution. "People still move to the United States, and that's largely
because of the . . . general understanding of science," he stated on
YouTube on August 23. "When you have a portion of the population
that doesn't believe in that, it holds everybody back."

In 2014, Nye held a debate against Ken Ham about evolution and
creationism at the Creation Museum in Kentucky. Ham is the founder
and head of the Young Earth Creationist Ministry. His museum
features exhibits that take visitors through biblical history.

Millions of people watched the two-and-a-half hour debate on
television. No one won the debate—but countless people had opinions
to share about it. In 2015, Nye published his book *Undeniable:
Evolution and the Science of Creation*. The next year, he published

Nye often served as a voice of science for CNN. He challenged the network to provide more people of science to urge action on global warming, instead of airing the few who doubt such a thing exists.

Unstoppable: Harnessing Science to Change the World.

In 2013, Nye showed the world an entirely different side of him: as a dancer. "I love dancing," he told PBS. "In Seattle, I went dancing all the time." For years, Nye had wanted to appear on the television show *Dancing with the Stars.* Finally, he was invited. "I was very nervous," he admits. His opening dance number got more than three million views. His time on the show was cut short, however, when he fell and tore some ligaments in his knee.

"Disappointment is the key word," he told PBS. "I felt disappointed for the fans, like I had let them down and I couldn't be a part of the show anymore." He enjoyed the few performances he did, however. "It was a really great experience. If you ever get the chance—do it."

On a beautiful summer morning in 2017, millions of Americans came outside to watch a rare solar eclipse take place over the United States. Nye was one of them. He shared his thoughts on the cosmic event, honoring famous scientists of the past: "Copernicus, Newton, and Leavitt—who came to understand our solar system's planets and moons, who measured the fantastic distances between them and came to know their orbital motions," he stated to *CBS News.* "That we humble humans can understand all of this is remarkable, and despite all the troubles around us today, it fills me with optimism about our species and our future." He continued, "[The] eclipse will be a once-in-a-lifetime event. Delight in its beauty. But also, appreciate that our science got us here." He added, "I hope this brief period reminds us all that we share a common origin among the stars and that we are citizens of the same planet. . . . Let's celebrate being alive right now in this universe, and marvel at humankind's ability to observe . . . and to understand the cosmos and our place within it."

In 2017, Nye returned to television to host *Bill Nye Saves the World* on Netflix. This science show is geared for adults. It still includes his trademark humor, and he still reminds his viewers, "It's not magic—it's science!" The show features guests, experts, and hot science topics,

Two of Nye's biggest fans are astronomer Neil DeGrasse Tyson and former President Barack Obama.

Nye was one of the many participants in the 2017 March for Science.

such as vaccines, artificial intelligence, and climate change. He also published his book *Everything All at Once: How to Unleash Your Inner Nerds, Tap into Radical Curiosity, and Solve Any Problem.* In addition, he, along with author Gregory Mone, began publishing a series of books for younger readers. *Jack and the Geniuses* is a series about Jack and his foster siblings, Ava and Matt. Together the three of them—all geniuses, of course—use science to solve mysteries.

All his life, Nye has tried to share his passion for science with the world. He has this advice for young people: "You have got to be into something—you have to be passionate about something," he told the *LA Times.* "I like tinkering, I'm a tinkerer. So I became a mechanical engineer. But I can imagine people are nerdy about all sorts of things and we want you to take that passion and do something great with it. Save the world!"

He also reminds everyone that they have a responsibility to the Earth. "You know, humans now move more earth and rock than Mother Nature does. I mention that to point out that we have a responsibility. We are in charge now of the planet. So you can't just solve one problem or another problem. We have to solve all the problems and we have to solve them all at once."

Chances are that as people work to find these solutions, Bill Nye, the Science Guy, will be right there to help and answer questions.

1955   William Sanford Nye is born in Washington, D.C., on November 27.

1973   He graduates from Sidwell Friends (High) School.

1977   He graduates from Cornell University with a degree in mechanical engineering.

1990   He appears on the comedy sketch show *Almost Live!* in Seattle, as well as on KJR Radio.

1991–1993 He performs in live-action segments on *Back to the Future: The Animated Series.*

1993   The first episode of *Bill Nye, the Science Guy* airs on September 10.

1998   The last episode of *Bill Nye, the Science Guy* airs on June 20; Nye appears in the Disney movie *The Principal Takes a Holiday.*

2002–2003 He works as technical expert for the TV show *BattleBots.*

2004   Nye helps create the MarsDials sent on the Rover probes to Mars.

2004-05   He hosts the eight-part miniseries on Discovery Channel called *Greatest Inventions.*

2005   He hosts the TV show *Eyes of Nye.*

2005–2010 Nye is executive director of the Planetary Society.

2010   He hosts the permanent Climate Lab Exhibit at Chabot Space & Science Center in Oakland, California.

2012   Nye makes headlines for challenging those in denial of evolution.

2013   He is a contestant on *Dancing with the Stars*.

2014   He debates evolution vs. creationism with Ken Ham at the Creation Museum in Kentucky.

2015   Nye publishes *Undeniable: Evolution and the Science of Creation*.

2016   He publishes the follow-up book, *Unstoppable: Harnessing Science to Change the World*.

2017   Nye publishes *Everything All at Once*; *Bill Nye Saves the World* premieres on Netflix; with Gregory Mone, he publishes the first two books in the *Jack and the Geniuses* chapter book series.

Books

Nye, Bill. *Bill Nye the Science Guy's Big Blast of Science*. Mercer Island, Washington: Perseus Books, 1993.

Nye, Bill, and Gregory Mone. *Jack and the Geniuses: At the Bottom of the World*. New York: Amulet Books, 2017.

Nye, Bill, and Gregory Mone. *Jack and the Geniuses: In the Deep Blue Sea*. New York: Amulet Books, 2017.

Nye, Bill, and Ian Saunders. *Bill Nye the Science Guy's Consider the Following: A Way Cool Set of Science Questions, Answers, and Ideas to Ponder*. New York: Scholastic Books, 1996.

Works Consulted

"12 Things You Didn't Know about Bill Nye, the Ultimate Guy." *Ultiworld*. August 26, 2014. https://ultiworld.com/2014/08/26/12-things-didnt-know-bill-nye-ultimate-guy/

"Bill Nye." *Famous Scientists*. Undated. https://www.famousscientists.org/bill-nye/

"Bill Nye on an Eclipse's Beauty and the Wonder of Science." CBS News. August 20, 2017. https://www.cbsnews.com/news/bill-nye-on-an-eclipses-beauty-and-the-wonder-of-science/

"Bill Nye, the Sundial Guy." *Astrobiology*. October 8, 2003. http://www.astrobio.net/mars/interview-with-bill-nye-the-sundial-guy/Booth, Tiina.

Dyak, Brian. "Science Guy Bill Nye Explores Life's Meaning in Five Minutes." *Live Science*. April 9, 2014. https://www.livescience.com/44699-science-guy-bill-nye-on-life.html

Friend, Tad. "The Sun on Mars." *The New Yorker*. January 5, 2004. http://www.newyorker.com/magazine/2004/01/05/the-sun-on-mars

Haner, Jim. "Jacqueline Jenkins-Nye, 79, World War II Code Breaker." *The Baltimore Sun*, April 3, 2000. http://articles.baltimoresun.com/2000-04-03/news/0004030047_1_bill-nye-goucher-college-war-ii

Kessler, Sarah. "How Bill Nye Became the Science Guy. And a Ballet Shoe Inventor. And a Political Voice." *Fast Company*, October 1, 2012. https://www.fastcompany.com/3001653/how-bill-nye-became-science-guy-and-ballet-shoe-inventor-and-political-voice

"Know Your Scientist: Bill Nye, "The Science Guy." *Futurism*. November 6, 2013. https://futurism.com/know-your-scientist-bill-nye-the-science-guy-2/

Koperski, Scott. "Bill Nye 'The Science Guy' Draws Crowd. *Beatrice Daily Sun*. August 21, 2017. http://beatricedailysun.com/news/local/bill-nye-the-science-guy-draws-crowd/article_e604e670-5439-51c3-883e-40946eb72ad9.html

Nye, Bill. "The Secret Life of Scientists and Engineers." PBS. Undated. http://www.pbs.org/wgbh/nova/blogs/secretlife/engineering/bill-nye/

Pastore, Rose. "8 Awesome Things We Learned about Bill Nye from His Reddit AMA." *Popular Science*. June 5, 2013. http://www.popsci.com/science/article/2013-06/8-awesome-things-we-learned-about-bill-nye-his-reddit-ama

Schwarze, Kelly. "How Bill Nye's Astronomy Teacher, Carl Sagan, Changed His World." *Popsugar*. March 23, 2014. https://www.popsugar.com/tech/Carl-Sagan-Influence-Bill-Nye-34412414

On the Internet

Bill Nye: Episode Guide
http://www.tv.com/shows/bill-nye-the-science-guy/episodes/
Bill Nye: Netflix Guide
https://www.netflix.com/title/80046944
Bill Nye Official Website
http://billnye.com/
Disney: Bill Nye
http://dep.disney.go.com/billnye.html

astrophysicist (ast-roh-FIH-zih-sist)—A person who studies the physical and chemical properties of the stars, planets, and other objects in outer space.

codebreaker (KOHD-bray-ker)—A person who solves secret codes to see what they mean.

controversial (kon-truh-VER-shul)—Causing a great deal of discussion, disagreement, or argument.

creationism (kree-AY-shun-ism)—The theory that the world was created exactly the way it is described in the Old Testament of the Bible.

device (dee-VYS)—An object, machine, or piece of equipment that has been made for a special purpose.

Emmy (EH-mee)—An award that is given each year to the top television shows and the people who work on them.

evolution (eh-vuh-LOO-shun)—The theory that the differences among modern plants and animals are the result of changes caused by natural processes over a very long time.

ligament (LIG-uh-mint)—A strong piece of tissue in the body that holds bones together or keeps organs in place.

literate (LIH-ter-it)—Able to read; knowledgeable about a subject.

orbital (OR-bih-tul)— relating to the curved path that a moon or satellite follows as it goes around a larger object in space.

patent (PAT-ent)—A document that gives an inventor sole rights to an invention.

physics (FIH-ziks)—Science that deals with matter and energy and the way they act on each other.

probe (PROHB)—An instrument used to get information from outer space and send it back to Earth.

quartermaster (KWOR-ter-mas-ter)—An army officer who provides clothing and other supplies for soldiers.

scholarship (SKAH-lur-ship)—Money given to students to help them pay for education.

solar eclipse (SOH-lur ee-KLIPS)—When the moon passes between the sun and Earth, so that the sun looks as if it is covered by the moon.

PHOTO CREDITS: Cover—Montclair Film; p. I—Simon Fraser University; pp. 4, 18, 20, 21—NASA; p. 8—Gildardo Sanchez; p. 12—Meutia Chaerani; p. 15—Dylan Otto; p. 17—Reynaldo Leal; p. 22—Neil Grabowski; p. 24—Jim Trottier; p. 26—Becker19. All other photos—public domain. Every measure has been taken to find all copyright holders of material used in this book. In the event any mistakes or omissions have happened within, attempts to correct them will be made in future editions of the book.